Carlos Calvet

Gravity Manipulation and its Binary Nature

GRIN Verlag

Bibliografische Information der Deutschen Nationalbibliothek:

Die Deutsche Bibliothek verzeichnet diese Publikation in der Deutschen Nationalbibliografie; detaillierte bibliografische Daten sind im Internet über http://dnb.d-nb.de/ abrufbar.

Imprint:

Druck und Bindung: Books on Demand GmbH, Norderstedt Germany
ISBN: 978-3-638-65494-4

This book at GRIN:

http://www.grin.com/en/e-book/39804/gravity-manipulation-and-its-binary-nature

PART 1

Gravitation and Inertia as a Consequence of Quantum Vacuum Energy

Carlos Calvet

Abstract: By assigning the elementary Planck units to the units of Newton's Gravitational Constant (G), it resulted in G being a function of vacuum (zero point) energy (ZPE). ZPE appears to reduce gravity, as it is inversely proportional to gravitational force. Further, the value of ZPE density-matter equivalent has resulted to be equivalent to the Planck mass in a Planck volume, rendering a much easier way of calculation.

Key words: Gravity, Gravitational constant, zero point energy, inertia, electrogravity, quantum vacuum.

Introduction

It was Isaac Newton, who 1687 found first the laws of motion and gravitation. He observed that two masses attract mutually, with a force that is directly proportional to the product of the masses and inversely proportional to the square of the distance. The resulting attractive force was in addition always a multiple of this proportionality, being the corresponding constant G (Gravitational Constant), which has an almost constant value of $6.673x10^{-11}$ $m^3kg^{-1}s^{-2}$. It has been up to date generally accepted that this "natural" constant is of unknown origin.

More than 2 centuries later, Max Planck discovered in 1900 that light quanta could explain black body radiation, and thus developed his black body law and his constant. Planck's constant (*h*) has the value $6{,}626x10^{-34}$ J·s, and represents the smallest energy amount that can exist, demonstrating the real existence of light quanta and overall quantification.

Planck's constant was the starting point for the calculation of some natural units for length, time and mass. Planck showed, simply based on a comparison of units, that by means of G, the speed of light (*c*) and his constant (*h*), it is possible to calculate an elementary length, time, and mass, which is now known as Planck's mass, length, and time (m_P, l_P, t_P). The currently accepted values for these Planck units are respectively $2.177x10^{-8}$ kg, $1.616x10^{-35}$ m, and $5.391x10^{-44}$ s, with the latter being in respect to General Relativity, 'the smallest length and time, spacetime can sustain'. Intriguing for our purposes was that Planck's units are a function of G (i.e., $m_P = (h/(2\pi)\ c/G)^{1/2}$, $l_P = (h/(2\pi)\ G/c^3)^{1/2}$, $t_P = (h/(2\pi)\ G/c^5)^{1/2}$). This obviously suggested already a century ago, that G is a quantum function.

In 1926, Werner Heisenberg developed the Uncertainty Principle (UP), which was the starting point of a new interpretation of absolute vacuum. According to the UP, a vacuum cannot be completely empty, i.e., it ought to display some background activity in order to allow its own existence. One year later, Paul Dirac described the quantification of electromagnetic fields, creation and anihilation of pairs, virtual particles, and ZPE, suggesting for the first time that an active "quantum vacuum" (QV) really exists.

On the other hand, H.G.B. Casimir [1], from Dutch Philips Laboratories, discovered in 1948 an attractive force between two very close 'perfectly conducting plates', which was opposite to the repulsive electric effect of the plates. The force was confirmed and measured precisely by Lamoreux [2]. Milonni et al. [3] showed furthermore that Casimir-plates are being pushed together by the 'unbalanced ZPE radiation pressure'. It seems that this effect derives from the partial shielding of the interior region of the plates from the background ZP fluctuations of the vacuum EM (electromagnetic) field, suggesting again the real existence of a QV and, in addition, of vacuum radiation. Further, B. Haisch, A. Rueda, and H.E. Puthoff deduced for the first time that the "inertia of matter could be interpreted at least in part as a reaction force originating in interactions between the EM ZPF and the elementary charged constituents (quarks and electrons) of matter" [4]. They also accepted that "extensions to include the ZPFs of other fundamental interactions may be necessary for a complete theory of inertia", which suggests in a wider extent that, in addition to Newton's equation of motion (F=m·a), there could be other parameters suitable to being redefined as ZPE-functions [4]. In recent publications [5], [6], H. E. Puthoff and B. Haisch, and A. Rueda calculated respectively the "mass-density equivalent of the vacuum ZPF fields" (10^{97} kg/m^3) and the "maximum energy density, spacetime can sustain" (10^{115} J·m-^{3}s^{-1}).

The present paper combines Planck values and the mass-density equivalent of the vacuum ZPF fields with gravity, providing a yet unknown equation, which demonstrates that *G is a function of vacuum energy* and that *this energy weakens "mass attraction"*. Furthermore, new equations for inertia are given and 'electrogravity' in 'weak gravity shielding experiments' is explained as the result of ZPE-manipulation through EM devices.

1. *"G" as a Function of QV-ZPF Mass-Density Equivalent.*

Newton's law of gravitation:

$$F = G\frac{m_1\, m_2}{d^2} \quad , \qquad (1)$$

contains the "natural" constant (G), with almost constant value of 6.673x10^{11} m^{3}kg^{-1}s^{-2}. By trying to reveal the meaning of units of G (m^{3}kg^{-1}s^{-2}), we found that they can be best represented by the inverse of "mass-density" (kg/m^3) multiplied by the inverse of "square time". In this sense, G can be expressed by the following equation:

$$G = \frac{1}{\delta}\frac{1}{t^2} \quad , \qquad (2)$$

where: δ = mass-density (kg/m^3), and: t = "a certain time" (s).

The following step was to assign concrete values to δ and t, in order to get the most exact value of G possible. Since G is a "universal constant", the most likely is that both, δ and t, are themselves constants, so that a medium corresponding to these values had to be found, that is constant in space and time of the whole universe. The only medium that fulfills this condition results to be QV, since it is per definition the most universal medium possible. In addition, the value of δ had already been calculated by [5] as the "mass-density equivalent of the vacuum ZPE fields" (10^{97} kg/m^3).

In this sense, finding t in equation (2):

$$t = \left(\frac{1}{\delta}\frac{1}{G}\right)^{1/2} \quad , \qquad (3)$$

and substituting G and δ by their respective values (all values in MKS-metric system), we get:

$$t = \left(\frac{1}{10^{97}\, kgm^{-3}}\,\frac{1}{6.673x10^{-11}\, m^3kg^{-1}s^{-2}}\right)^{1/2} = 3.871x10^{-44}\, s \;, \quad (4)$$

with the result that, "t" resembles much Planck Time ($t_P = 5.391x10^{-44}$ s).

Since there is no other "time" that resembles t_P so closely, and provided that "t" is effectively "t_P", it is legitimate to express G as:

$$\boxed{G = \frac{1}{\delta_{ZP}\ t_P^2}} \quad , \qquad (5)$$

where: δ_{ZP} = ZPF mass-density equivalent.

Finding now δ_{ZP} in (5), we are able to calculate the exact value of the ZPF mass-density equivalent:

$$\boxed{\delta_{ZP} = \frac{1}{G\ t_P^2} = 5.156\,x10^{96}\ kg/m^3} \quad , \qquad (6)$$

which is almost identical to the10^{97} kg/m^3 that Puthoff [5] calculated for the "mass-density equivalent of the vacuum ZPE fields".

2. *Demonstration that (5) is Correct.*

Since the figure of δ_{ZP} [5] was an approximate value, this author developed an alternative way to demonstrate that (5) is correct. In fact, a QV mass-density can be understood per definition as a Planck mass in a Planck volume:

$$\boxed{\delta_{ZP} = \frac{m_P}{l_P^3}} = \frac{2.177x10^{-8}kg}{\left(1.616x10^{-35}m\right)^3} = \frac{2.177x10^{-8}\,kg}{4.220x10^{-105}m^3} = 5.159x10^{96}\,kgm^{-3}, \quad (7)$$

with the values of (6) and (7) being identical to the rounded decimals. The corresponding mean is $5.1575x10^{96}$ kg/m^3 and in any case, δ_{ZP} is equal to rounded $5.16x10^{96}$ kg/m^3. Since (6) and (7) are practically identical, it is legitimate to consider that (5) is a correct equation in describing G.

3. *Relationship between the ZPF Mass-Density Equivalent and ZPE Density Flow.*

Taking Haisch & Rueda's [6] equation of the ZPF "energy density" flow at the Planck frequency cutoff ($\rho_{ZP} = 2\pi^2c^7/G^2\,h$) and finding G^2, we get:

$$G^2 = \frac{2\pi^2\,c^7}{\rho_{ZP}\,\hbar} \quad . \qquad (8)$$

On the other hand, taking equation (5) and substituting t_P by its value ($h/(2\pi)\ G/c^5)^{1/2}$), we can find G^2:

$$G = \frac{1}{\delta_{ZP}\, t_P^2} = \frac{1}{\delta_{ZP} \frac{\hbar}{2\pi} \frac{G}{c^5}} \Rightarrow G^2 = \frac{2\pi\, c^5}{\delta_{ZP}\, \hbar} \quad . \qquad (9)$$

Now, equaling equations (8) and (9), we get

$$\frac{2\pi\, c^5}{\delta_{ZP}\, \hbar} = \frac{2\pi^2\, c^7}{\rho_{ZP}\, \hbar} \quad , \qquad (10)$$

and in consequence

$$\boxed{\rho_{ZP} = \pi\, \delta_{ZP}\, c^2} \quad , \qquad (11)$$

which means that QV energy flow is a function of QV energy density and the speed of light, thus unifying these two apparently different QV-parameters.

Conclusions and Discussion

1. *Gravitation as a Combined Force of Mass Attraction and QV Repulsion*

Through (5), Newton's equation of gravitation (1) can be now expressed more precisely (at quantum level) by the following equation:

$$\boxed{F = \frac{1}{\delta_{ZP}\, t_P^2} \frac{m_1\, m_2}{d^2}} \quad , \qquad (12)$$

where G has been substituted by a function of QV energy, and δ_{ZP} reflects the value of the "ZPF mass-density equivalent", which can be now calculated exactly through (6) and/or (7).

Since the left component of (12) is a constant, and the right a variable, they are mutually independent and can be treated separately. In consequence, the first we observe, while regarding equation (12), is that it is composed of two components: a QV (ZPE) component to the left, and a mass attraction component to the right. Secondly, δ_{ZP} is inversely proportional to gravitational attraction (F), what consequently means in principle that δ_{ZP} reduces gravity (while mass increases gravity). In consequence, we can call the right component "pure mass attraction" and the left, "vacuum repulsion reaction" since mass is obviously the inducing component of gravitation and QV probably reacts (as per the "cause-effect" principle) to the presence of mutually attracting masses.

In summary, we have found through equation (12) that gravitation is a combined force and that the QV-component (to the left) is opposed to the mutual gravitational attraction of masses (component to the right). The sense of this apparently complex gravitation-generating system in the universe is probably that of stabilizing gravitation, thus avoiding unbiased extreme mass-attractive forces, by opposing a constant reaction effect through QV. If there were no QV in the universe, G would be equal to 1 $m^3kg^{-1}s^{-2}$ (since it would mean that there is no QV reaction) and gravity would be much stronger than it is at present (exactly $6.673x10^{11}$ times stronger, since that would be the reverse value of G). In this case, the universe would probably shrink to a spot (Big Crunch), since there would be no effect able to stop a gravitational mass collapse.

2. *Gravitational Inertia*

Davies [7] and Unruh [8] demonstrated that there exists a real QV-reaction force to accelerated matter, such that mass acceleration and the opposed QV-reaction effect are two forces, which are intimately interrelated in nature, being the corresponding "Davies-Unruh effect" therefore apparently close to the definition of inertia (a reaction force to acceleration). [7] and [8] demonstrated to a wider extent the link existing between masses and QV.

Since in the macroscopic world, inertia is the immediate reaction effect opposed to acceleration, the left component of eq. (12) is therefore analogous to inertia as it is the reaction force to mass acceleration. In addition, following the above-mentioned "principle of independency" and treating the left and the right components of eq. (12) independently, gravitation can be redefined as consisting of two components:

$$\boxed{F = F_M - F_{ZP}} \quad , \qquad (13)$$

where F_M is a 'pure mass attraction force' due to the purely attractive gravitational effect between two masses as if there were no QV (derived from eq. (12) when G = 1 $m^3kg^{-1}s^{-2}$, i.e., G has no effect); F_{ZP} is the corresponding decelerating ZPF-reaction force, macroscopically known as *inertia*; and F is the resulting natural attraction force, which can be measured by man, where G has the known natural value of $6.673x10^{-11}$ $m^3kg^{-1}s^{-2}$.

Finding *inertia* in (13):

$$\boxed{F_{ZP} = F_M - F} \quad , \qquad (14)$$

and substituting the above mentioned values of G in (14), while expressing each force in the sense of Newton's gravitation law, we get the value for *gravitational inertia*:

$$\boxed{F_{ZP} = ((1 - 6.673\,x10^{-11})\,m^3kg^{-1}s^{-2})\,\frac{m_1\,m_2}{d^2}} \quad . \qquad (15)$$

As seen in eq. (15), the value of G for *gravitational inertia* ($(1 - 6.673x10^{-11})$ $m^3kg^{-1}s^{-2}$) is very close to 1 (0.99999999993327 $m^3kg^{-1}s^{-2}$) and demonstrates that inertia is a very strong force as expected from its obviously strong decelerating effect on matter in opposition to light (and other photons) that is commonly not decelerated by inertia or by any other similar effect.

Since inertia depends on G, and because according to (5), G is a QV-function, if QV did not exist, there would be no gravitational inertia in the universe, and celestial bodies and masses in general would attract mutually without any control, probably shrinking extremely, since an unbiased mass attraction would possibly make any mass in the universe collapse into small bodies like neutron stars or black holes. QV seems to be therefore a "background" that guarantees the structural stability of our universe and should furthermore not be manipulated in a global extend without stringent physical control.

3. *Non-Gravitational Inertia*

For non-mutually attracting bodies, i.e., for individual accelerated objects, (non-gravitational) inertia can be derived from (14), by substituting forces through Newton's equation of force (F=ma). For a certain mass, pure mass attraction (F_M) is per definition equal to the force, it would produce in the case of the highest acceleration possible, that is, the Planck acceleration ($a_P=l_P/t^2_P$ = 5.560x10^{51} m/s^2). In consequence, substituting a_P in (14) in the sense of Newton's equation of force, we get:

$$F_{ZP} = m\,a_P - m\,a = m(a_P - a) = m(5.560\,x10^{51} ms^{-2} - a) \quad . \qquad (16)$$

This is inertia of an independent accelerated object of mass *m* and acceleration *a*, and shows that inertia of non-gravitational acceleration too limits strongly the independent mass acceleration, as already expected with regard to (15). Eq. (16) predicts correctly the zero inertia of a photon, since in this case, "a" would be equal to a_P (being $a_P = c/t_P$, where c = speed of a photon) and F_{ZP} = 0.

In consequence, if QV did not exist, there would be no reaction force to non-gravitational acceleration and bodies would reach unbiased velocities in the universe, which probably would not allow the formation of stable celestial bodies as they were always destroyed by heavy collisions with other bodies that would have been violently accelerated by some forces.

In general, without the stabilizing effect of QV, there would be no inertia at all and the universe would obviously be a very chaotic place, where matter collided without any control at very high speeds and unbiased gravitational forces favored the formation of very massive bodies that would carry very high collision energies, thus rendering an almost self-destructive and/or crunching universe that probably could not even exist for more than a single moment.

4. *Gravity Control through Electromagnetism*

Since G has been proven to be a QV-function by (5), the same applies to gravitation through (12), thus providing the realistic possibility of gravity control through manipulation of ZPE. In fact, (12) demonstrates that QV-energy density weakens gravity. In consequence, if we were able to manipulate ZPE, we would be altering gravity through (12). By increasing QV-energy density, gravity would decrease, while by decreasing the QV-energy density, gravity would increase.

Podkletnov [9] discovered in this sense, in a very controversial work, that a "composite bulk YBa2Cu3O7-x superconductor below 70°K under EM field" was able to produce, what he called 'weak gravitation shielding', above and below his superconductor arrangement. The experiment was reproduced by Li et al. [10] and others, and explained by this team, Modanese [11], and others. According to [10], "rotating superconductors in an alternating magnetic field would generate gravity". NASA is studying this effect in its High Temperature Superconductor (HTSC) Research Program, with an aim towards developing technologies for future interstellar navigation.

According to (12), gravity weakens if ZPE increases. Therefore, to produce a 'gravity shielding effect', the above arrangement should have been able to increase local ZPE density. This could obviously have happened through the involved magnetic fields (superconductor, coils). In order for a magnetic field to be able to increase ZPE, it is necessary that a transfer of photons from the magnetic fields to the ZPF, takes place. If this happened, then the higher concentration of vacuum radiation (photons) around the arrangement would produce a higher radiation pressure on nearby objects, thus lowering their weight as predicted by (12), what was effectively observed by Podkletnov.

In consequence, as (12) shows how gravity could be controlled through ZPE manipulation, Podkletnov's experiment seems to suggest strongly that this happened through EM fields. Therefore, so-called "electrogravity" (the unification of gravitation and EM fields) could be achieved directly by combining (12) and electromagnetism in arrangements similar to Podkletnov's. The driving force that altered gravity would be in the most-simple case, the same well-known vacuum radiation detected in the Casimir-effect [1].

Finally, the main conclusion of this paper is that quantum vacuum is essential for the existence of spacetime, so that we cannot imagine spacetime without the effects of vacuum radiation. The inclusion of QV in physics provides much clearer equations that tend to simplify each other, even down to a yet unknown extend, with the possibility to get very close to the so-called "universal formula". The corresponding much better understanding of the physical world may lead us to unexpected technologies that arise at Planck level such as gravity manipulation, dimensional control, "stargate" technology, string-colliders etc. [12].

References

[1] H.G.B. Casimir, "On the attraction between two perfectly conducting plates", Proc. Kon. Ned. Akad. van Weten., Vol. 51, No. 7, pp. 793-796 (1948).

[2] S.K. Lamoreux, "Demonstration of the Casimir force in the 0.6 to 6 mm range", Phys. Rev. Lett., Vol. 78, No. 1, pp. 5-8 (1997).

[3] P.W. Milonni, R.J. Cook, & M.E. Goggin, "Radiation pressure from the vacuum: Physical interpretation of the Casimir force", Phys. Rev. A, Vol. 38, No. 3, pp. 1621-1623 (1988).

[4] B. Haisch, A. Rueda & H.E. Puthoff, "Inertia as a zero-point-field Lorenz force", Physical Review A, Vol. 49, No. 2, pp. 678-694 (1994).

[5] H.E. Puthoff, "The Energetic Vacuum: Implications For Energy Research, Speculations in Science and Technology", Vol. 13, No. 4, pp. 247-257 (1990).

[6] B. Haisch, A. Rueda, "Toward an Interstellar Mission: Zeroing in on the Zero-Point-Field Inertia Resonance", AIP Conference Proceedings of the Space Technology and Applications Forum (STAIF-2000) Conference on Enabling Technology and Required Scientific Developments for Interstellar Missions, January 30-February 3, Albuquerque, NM (2000).

[7] P.C.W. Davies, "Scalar particle production in Schwarzschild and Rindler metrics", J. Phys. A, Vol. 8, p. 609 (1975).

[8] W.G. Unruh, "Notes on black-hole evaporation", Phys. Rev. D, Vol. 21, p. 2137 (1980).

[9] E.E. Podkletnov (Moscow Chem. Scientific. Ctr.), "Weak Gravitation Shielding Properties of Composite Bulk YBa2Cu3O7-x Superconductor Below 70°K under E.M. Field," Univ. Cincinnati Engineering, report # MSU-chem 95, abstract cond-mat/9701074, 19 pp. (1997).

[10] N. Li, D. Noever, T. Robertson, R. Koczor and W. Brantley, "Static Test for A Gravitational Force Coupled to Type II YBCO Superconductors", Physica C, Vol. 281, pp. 260-267 (1997).

[11] G. Modanese, "On the theoretical interpretation of E. Podkletnov's experiment", I.N.F.N. - Trento, Extract from report UTF-391/96, LANL gr-qc/9612022; presented for the World Congress of the International Astronautical Federation, 1997, No. IAA-97-4.1.07.

[12] Carlos Calvet, "About the quantum vacuum lepton/photon ratio", Journal of Theoretics, Vol. 4, No. 2, April 2002 (preprint)

PART 2

The Quantum Vacuum Lepton/Photon Ratio

Carlos Calvet planckworld2@terra.es
Francisco Corbera no. 15, E-08360 Canet de Mar (Barcelona), Spain
personal.telefonica.terra.es/web/planckworld/

Abstract: From Coulomb's Constant, the "Planck charge" (q_P) and the corresponding quantum vacuum (QV) lepton/photon ratio are calculated, the latter by dividing q_P by the Planck mass. The QV-lepton/photon ratio is very different to the analogous baryon/photon ratio that is predicted by Standard Model, demonstrating that QV and space-time are **two different entities** (e.g., that they have a *different number of dimensions*). A dimensional model for the QV that evolves from, and is supported by, the lepton/photon ratio is presented herein, together with emerging technologies.

Keywords: Coulomb's Constant, Planck charge, virtual pairs, quantum vacuum.

Introduction

That Coulomb's Constant ($C = 8.988 \times 10^{9}$ Nm^2/C^2) is a quantum-vacuum (QV) function, is already known, since it is a function of vacuum permittivity (e_0) and/or permeability (μ_0). That vacuum has a permittivity and/or permeability means that it contains charges, since any charge is exposed in vacuum to a certain electrical resistance. But no "real" charges are found in empty space, thus the corresponding particles (Dirac pairs) are considered "virtual".

In a former paper [1], this author found that Newton's Gravitational Constant ($G = 6.673 \times 10^{-11}$ $m^3kg^{-1}s^{-2}$) also is a QV-function, since it can be precisely derived from Zero Point Energy (ZPE) Density Matter Equivalent ($\delta_{ZP} = 5.159 \times 10^{96}$ kg/m^3) and Planck time (t_P) according to the equation: $G = \delta^{-1}_{ZP}\ t^{-2}_{P}$.

In the same paper, it was demonstrated that the value of the ZPE Density Matter Equivalent found (e.g. in [2] and [3]; value of approx. 10^{97} kg/m^3) is much easier obtained, by simply dividing Planck mass (m_P) by Planck volume (third power of Planck length [l_P], being $m_P/l^3_P = 5.159 \times 10^{96}$ kg/m^3), with the corresponding energy being as "virtual" as QV-charges (i.e., it is unable to be detected directly).

Further, it is generally known that even the speed of light (c) is a QV-function, as it is the exact quotient between l_P and t_P ($c = l_P/t_P$). This fact and that the above-mentioned ZPE Density Matter Equivalent (DME) is equal to the quotient between Planck mass and Planck volume, supports the idea that Planck units are real and that they represent effectively existing physical circumstances of QV.

On the other hand, natural constants like Coulomb's Constant and Newton's Gravitational Constant are per definition absolutely real too, since they represent space parameters that influence our measurements, respectively by increasing or decreasing natural forces, in this sense, electric force ($C > 1$ Nm^2/C^2) and gravity ($G < 1$ $m^3kg^{-1}s^{-2}$).

When Max Planck derived in 1899 his natural units (l_P, t_P, m_P), he used a comparison method, whereby he took G, c, and his own constant (h), and thus assembled them according to the corresponding dimensions. In this way, he obtained a length, a time, and a mass that are as real as any natural constants, since they are the direct product of these. Even Planck frequency ($\nu_P = t_P^{-1}$) is a very real value, since it is the assumed cut-off of the QV-radiation spectrum [4] and this frequency was used in [3] to calculate the ZPE DME. In addition, in [1], it was demonstrated that the value of the ZPE DME is exactly the same, whether it is derived from ν_P or from m_P, thereby implying that the QV frequency cut-off is effectively ν_P and that both values are analogous and real.

Since ZPE is per definition the energy contained in QV, the most outstanding conclusion of the above-mentioned is that both, natural constants and Planck units, are real and provide direct evidence about the properties and composition of QV.

This paper follows the above-mentioned research lines and provides strong evidence that the QV and String Theory can be unified through a common 6-dimensonal space that allows well-known properties like particle non-locality or Bose-Einstein (B-E) condensate material waves.

Results

As demonstrated in [1], G is a function of the ZPE DME and, per extension, of the QV-energy content. Hereunder, it will be shown, how it is possible to calculate from Coulomb's Constant (C), the charge contained in QV, known as "virtual pairs".

Considering the units of C (Nm^2/C^2) and using the same method as Max Planck, when he developed his natural units (and per extension, the same method as used in [1]), we break down the units of C into Planck units as follows:

$$C = \frac{F\, l_P^2}{q^2} = \frac{E\, l_P}{q^2} = \frac{m_P\, c^2\, l_P}{q^2} \quad , \tag{1}$$

where: F = a force, E = an energy, and the value "q" represents in this context per definition, the charge that is confined within a Planck volume, therefore making it legitimate to denominate this, "Planck charge" (q_P).

Finding "q_P" and substituting the values of the other Planck units (all values hereunder being of the MKS metric system):

$$q_P = \left(\frac{m_P\, c^2\, l_P}{C}\right)^{1/2} = \left(\frac{2.177x10^{-8}\, kg\left(2.998x10^{8}\, ms^{-1}\right)^2 1.616x10^{-35} m}{8.988x10^{9}\, Nm^2C^{-2}}\right)^{1/2} = 1.876x10^{-18}\, C \ . \tag{2}$$

To convert the above q_P into the corresponding amount of leptons (where electrons and positrons = virtual pairs), we divide q_P by the charge of the electron ($1.6022x10^{-19}$ C), thus obtaining the amount of 11.71 leptons per Planck volume. These 11.71 leptons are then multiplied by the mass of the electron ($m_e = 9.1094x10^{-31}$ kg), corresponding to a *lepton mass equivalent* ($m_{q,P}$) of $1.0667x10^{-29}$ kg in a Planck volume.

Finally, since in the ZPE DME (m_P/l^3_P), m_P represents an electromagnetic (EM) radiation, by dividing $m_{q,P}$ through m_P, we obtain the QV-lepton/photon ratio:

$$\frac{m_{q,P}}{m_P} = \frac{1.0667x10^{-29}\, kg}{2.177x10^{-8}\, kg} = 4.9000x10^{-22} \quad . \tag{3}$$

Consequently, in QV, leptons (virtual pairs) are approximately only $4.9x10^{-20}$ % as frequent as photons (ZPE), what was already known qualitatively in vacuum research.

Conclusions and Discussion

It is generally known that in the universe, there are many more photons than fermions. The baryon/photon ratio as provided by the Standard Model is approximately 10^{-9}, although

Olive, Steigman, and Skillman [5] have argued that it is approximately 10^{-10}. Since, as generally accepted, the universe must have an almost equal number of positive and negative charges in order to maintain its own stability (as known, a slightly unbalanced charge ratio would make the universe collapse), it is legitimate to compare the lepton number directly with the baryon number, since both are equivalent.

In this sense, the QV-lepton/photon ratio (3) is about *12 orders of magnitude lower* than the value of the baryon/photon ratio in the Standard Model, demonstrating that in QV, the proportion of photons with respect to matter is much higher than in space-time (there have been found no neutral particles in QV that could alter this fact). The fact that charges are much less frequent in QV than in space-time with respect to photons, demonstrates that at least in principle, the "real" universe (space-time) and the "virtual" QV are two different spaces.

Since the parameters $m_{q,P}$ and m_P used in (3) represent, respectively, leptons/gamma particles and QV-radiation, and QV-radiation has its mean approximately at the level of such gamma particles (10^{20}-10^{22} Hz is approximately the mean of the whole EM QV-spectrum up to the Planck frequency, 1.855 x 10^{43} Hz), the ratio (3) is perfectly comparable to that of the Standard Model. And even if the ratio (3) should change slightly, this would have no effect because of the enormous difference observed (about 12 orders of magnitude).

The apparently trivial question of why the QV is 'virtual', appears to be extremely important if we consider that "at the 53 GHz-frequency middle band of the COBE Differential Microwave Radiometer, the ZPF/Cosmic Microwave Background ratio would be 0.77. Even more dramatically, in the optical spectrum, eqn. 2 (EM-blackbody spectrum including ZPF) predicts, that the ZPF should be about two orders of magnitude brighter than the Sun", as argued in [6].

The generally accepted explanation of why the QV is a virtual space, is that our universe is built upon QV-energy, such that we are not able to detect it, because any detector would have to be made out of that energy. Some known visible effects produced by the QV are so-called "van der Waal's forces", which are microscopic and act even near absolute zero (0° Kelvin); as well as adherence of crude Bose-Einstein condensate, and the macroscopic Casimir force between dielectric or conducting plates [6], [7]. But even so, the main question remains, "where exactly are ZPR (vacuum radiation) and virtual pairs located?" If they were located in 4-D space-time, we should be able to detect them directly, even if we and our devices were made of such energy.

Planck units do focus this problem. In fact, a Planck mass can be considered, per definition, as a certain natural QV-mass confined within a Planck volume and allows for the calculation of a QV-mass density equivalent of about 10^{97} kg/m^3 as mentioned in [1], [2], and [3]. If we divide m_P (2.177×10^{-8} kg) by the mass of an electron (9.110×10^{-31}kg), we get the corresponding concentration of 2.390×10^{22} leptons (electrons and positrons) per V_P (Planck volume). Even in the case of much heavier or more energetic particles (baryons, gamma-particles, etc.), we always get a concentration of particles that infringes on the principle that a Planck volume cannot contain more than *one* single particle in 4-D (dimensional) space-time, since it is per definition, the smallest possible volume that can exist in space-time.

One possible solution to the above problem could be, that the QV does not only contain photons and fermions, but also other bosons, like gluons. In this sense, [8] suggested that the QV might also be the source of fields of other fundamental interactions. But so far, only EM-radiation (ZPR) has been found to exist in QV.

The above-mentioned 11.71 leptons per Planck volume (V_P) mean that, on average, there are a little less than 12 leptons per V_P, so that most frequently, the integer number of leptons in a V_P is 12. Since these 12 leptons correspond to virtual pairs that anihilate and create mutually in an endless cycle together with the corresponding gamma-particles involved (virtual pairs produce gamma-particles and visa-versa), at any time, there is an amount of 6 pairs in a V_P, which at the very moment of their mutual interaction, take a space analogous to 6 particles or strings.

In this context, the California Institute for Physics and Astrophysics mentions on its homepage that "it now appears that quantum field theory may be the low energy limit of superstring theory". String theory predicts, as generally known, that strings represent a 6-D curled microscopic space in 4-D spacetime, so that the total dimensions of a particle is 10.

Finally, the 6 particle pairs found in a V_P are surprisingly coincident with the figure of 6 curled dimensions of strings, which correspond to the interior region of elementary particles. In addition, since space-time and QV are, as shown above, two *different spaces*, the immediate conclusion is that they also have a *different number of dimensions* (since two spaces with the same number of dimensions would belong obviously to the same space). But, as seen, QV and space-time have a different overall radiation content.

Taking the model of string-theory and applying it to QV, it results that QV is a 6-D space, since in this way, each QV-V_P would have just the necessary space to contain 6 particles (strings, equivalent to 6 virtual pairs at the very moment of their mutual interaction), so that each particle (string) would be correctly placed in its own and exclusive dimension. This is in principle possible, since strings are believed to be of the Planck length and one-dimensional. In consequence, any of the strings would be included in a Planck section of the corresponding dimension.

In consequence, the universe can be understood as consisting of a macroscopic 4-D space-time, surrounded by a macroscopic 6-D QV, being both of these two spaces linked together by elementary particles, the thermal surface of the particles being 4-D and interacting mainly with space-time, while the supercold interior region of the particles is 6-D and interacts mainly with the QV, being the total number of dimensions 10, as foreseen in string theory. Because of their above-mentioned apparently complex structure, particles can be understood as small interfaces that link space-time to QV.

The above model is very simple and able to explain novel phenomena like particle entanglement (quantum non-locality) and B-E condensation. In this sense, particle entanglement is interpreted by this model as particle communication between the interior 6-D regions of two or more particles through a 6-D macroscopic QV. The time needed for this kind of communication has obviously no meaning in 4-D space-time (it happens beyond 4-D spacetime), so that it is apparently zero for human observers, thus producing the well-known properties of entangled particles (non-locality, etc.).

B-E condensation strangeness too can be explained perfectly by this model, since at condensation temperature (±0K), the 4-D kinetic of the outer thermal shell of elementary particles becomes so weak, such that the interior 6-D quantified supercold region emerges and becomes visible to space-time-observers, thus providing quantum properties to B-E-condensed matter.

Since the QV and space-time are two incompatible spaces (i.e., they cannot merge mutually due to the different radiation contents), nature has obviously provided a so-called 'event horizon' (already known from black holes), which represents the interface of contact between 4-D space-time and the 6-D QV. This interface becomes evident, for example in the *wavy nature* of B-E-condensates, since it represents a quantum property emerged from the QV that is visible for space-time observers. The very slow speed of light observed in B-E condensates is probably the effect of such an event horizon on light, such that it travels normally through the QV, but seems to 'creep', viewed from space-time.

Since B-E condensates represent an event horizon between space-time and the QV, it should be possible to send signals through the QV to distant places in almost zero human time, and to develop new related technologies. Current B-E condensates are simple atom clouds, while superfluids (supercooled helium) are still very contaminated with thermal matter. But, if we managed to produce supersolids (supercold solid matter) or perfect superfluids, as soon as the whole block of matter became a wave, this wave would represent a *macroscopic interface* to the 'other side', opposite to the *microscopic interface* that particles represent. In this case, it ought to be possible, even to step through the matter wave and send, for example, small probes to explore the 'parallel universe' on the QV-side (with the necessary technology provided).

On the other hand, *string colliders* seem to be in reach, by making collide supercold atoms mutually inside energy-free "Casimir-spaces". Since the thermal 4-D surface that surrounds particles in spacetime weakens and disappears almost completely in B/E-condensates, the amount of energy needed to make collide the free strings of the interior region of such supercold and "nacked" particles will be much less than at higher temperatures. We predict therefore that string colliders will render a *huge energy gain* in comparison to other common methods of energy production.

Summary

This model agrees with the Big-Bang theory and complements it, since the QV and the interior region of particles, which have the same number of dimensions (6), will have probably had the same creation process too. As seen, one function of the QV in our universe is that of providing gravitation through vacuum radiation [1], [2]. If the QV did not exist, there would further be no particles, since their 6-D strings could not exist in 4-D space-time, so that, although the QV does not contain matter, rather it allows for a material existence in space-time. The supercold quantified 6-D condition of the QV probably hides critical events to future exploration. Nevertheless, efforts should be made to reveal the QV-nature through supercold matter experimentation, and subsequently develop QV-technology by using the above-mentioned event horizon szenario. A more detailed description of all these emerging technologies can be found in [9].

References

[1] C. Calvet, “Gravitation and Inertia as a Consequence of Quantum Vacuum Energy”, Journal of Theoretics (preprint).

[2] H. E. Puthoff, "Gravity as a Zero-Point-Fluctuation Force," Phys. Rev. A 39, 2333 (1989); Phys. Rev. A 47, 3454 (1993).

[3] B. Haisch, A. Rueda, “Toward an Interstellar Mission: Zeroing in on the Zero-Point-Field Inertia Resonance”, AIP Conference Proceedings of the Space Technology and Applications Forum (STAIF-2000) Conference on Enabling Technology and Required Scientific Developments for Interstellar Missions, January 30-February 3, Albuquerque, NM (2000).

[4] H.E. Puthoff, “The Energetic Vacuum: Implications For Energy Research, Speculations in Science and Technology”, Vol. 13, No. 4, pp. 247-257 (1990).

[5] Olive, K.A., Steigman, G. & Skillman, E., ‘‘The Primordial Abundance of 4 He: An Update,’’ ApJ, N° 483, pp. 788 (1997).

[6] B. Haisch, A. Rueda & H.E. Puthoff, “Physics of the zero-point field: implications for inertia, gravitation and mass”, Speculations in Science and Technology, Vol. 20, pp. 99-114 (1997) (preprint).

[7] H.G.B. Casimir, “On the attraction between two perfectly conducting plates”, Proc. Kon. Ned. Akad. van Weten., Vol. 51, No. 7, pp. 793-796 (1948).

[8] A. Rueda, B. Haisch, “Inertia as Reaction of the Vacuum to Accelerated Motion”, Phys. Let. A, Vol. 240, No. 3, p. 115 (1998).

[9] Carlos Calvet, “Principles of gravity manipulation and “Stargate”-technology via Quantum Vacuum”, Journal of Theoretics, Vol. 4, No. 4, August 2002 (preprint).

PART 3

Principles of Gravity Manipulation via the Quantum Vacuum

Carlos Calvet planckworld2@terra.es
Francisco Corbera no. 15, E-08360 Canet de Mar (Barcelona), Spain
personal.telefonica.terra.es/web/planckworld/

Abstract: By expressing natural constants in terms of Planck units, we found that the Universal Gravitation Constant is the inverse of vacuum density matter-equivalent and the square of Planck time, being the former equal to Planck mass divided by Planck volume. The corresponding new equation of gravitation presented here reveals that gravitation can be manipulated via vacuum energy. Additionally, from Coulomb's constant, we can derive the "Planck charge" and the corresponding density of virtual vacuum particle pairs. A discussion of the ramifications of these findings is also presented.
Key words: Gravitation, zero point energy, inertia, electrogravity, Coulomb's Constant, Planck charge, virtual pairs, quantum vacuum, event horizon.

Introduction

In two articles [1], [2], we derived respectively the Universal Gravitation Constant (G) from Planck units and demonstrated that, spacetime and quantum vacuum (QV), are two different spaces. Newton's equation of gravitation has two components (a constant [G] and a variable mass component [m_1m_2/d^2]) that can be treated separately, since they are independent. As a consequence, gravitation can be considered a combined force, consisting of G (as previously demonstrated to be a QV function) and conventional 'mass attraction', the latter produced by gravity fields and/or spacetime geometry according to the corresponding theories. This principle of independence has allowed us to calculate inertia for mutually attracting and/or otherwise accelerated bodies, finding that inertia has a very high value. Its fundamental effect is probably that of marking a clear distinction between fermions (matter) and photons (light), the latter not being affected by inertia.

While "mass attraction" depends only on mutually attracting masses and their relative position or distance, G (a so-called, "non-derivable constant") can be effectively derived and is found to be the exact inverse of "vacuum mass density equivalent" (5.156×10^{96} kg/m^3), which is come to by dividing the Planck mass by the product of Planck volume and the square of Planck time (the former having already been predicted as an approximate value, e.g. in [3]). This new equation of gravitation resulting from the substitution of G by Planck units reveals that vacuum density (analogous to Zero Point Radiation [ZPR]) affects gravity inversely. This means that, if we were able to increase ZPR, gravity would decrease and vice-versa. In fact, the very small value of G (6.673×10^{-11} m^{3}kg^{-1}s^{-2}) already suggests that there is 'something' that weakens mass attraction. We found in [1] that this 'something' is ZPR).

The above findings, in combination with so-called "weak gravitation shielding experiments" ([4], [5]), reveal that gravity can be effectively manipulated via QV by using superconductor arrangements.

In our second article [2], we derived the 'Planck charge' (generally unknown or as the 'electron charge' misunderstood with a value of 1.876×10^{-18} C) from Coulomb's Constant (C), corresponding to the charge existing in a QV-Planck volume. By comparing the corresponding QV lepton/photon ratio with the almost equivalent baryon/photon ratio predicted for spacetime by the Standard Model, we found that the former is about 12 orders of magnitude lower, allowing it to be said that under any circumstance, spacetime and QV are two different spaces (even if we corrected somehow for the corresponding ratios, the difference would be still too large to be merely insignificant).

These findings allow us to determine that spacetime is mainly a space full of neutral matter and much light, while the QV is mainly a space full of strong radiation and some charges (virtual pairs), such that both universes are effectively incompatible and ought to exist therefore - even from a theoretical point of view - as separated spaces in the universe.

By dividing the Planck charge by the charge of the electron ($1.6022x10^{-19}$ C), we obtain an average density of 11.71 leptons per Planck volume, which per definition corresponds to the well-known virtual electron-positron pairs that create and mutually annihilate each other in vacuum. At the very moment of their mutual interaction, approximately 12 entire leptons correspond to 6 particles or strings, what is coincident with the 6 Kaluza-Klein dimensions, attributed to strings.

This coincidence allows the establishment of a direct link between strings and the QV as already suggested by The California Institute for Physics and Astrophysics in its homepage ('It now appears that quantum field theory may be the low energy limit of superstring theory") and suggests that the QV is the space where strings are physically located. This model would explain the effect known as 'quantum non-locality', since strings (in so-called 'entangled particles') would interact instantly via the QV in a time that is zero for spacetime observers, since the QV disposes of a different time frame than spacetime because of its different number of dimensions, and therefore QV-time has no meaning to us.

Gravity Manipulation

As already shown in [1], vacuum mass-density equivalent can be understood, per definition, as a Planck mass existing in a Planck volume:

$$\boxed{\delta_{ZP} = \frac{m_P}{l_P^3}} = \frac{2.177x10^{-8}kg}{\left(1.616x10^{-35}m\right)^3} = \frac{2.177x10^{-8}kg}{4.220x10^{-105}m^3} = 5.159x10^{96}kgm^{-3} \quad , \qquad (1)$$

where: δ_{ZP} = vacuum mass-density equivalent, m_P = Planck mass, and $l^3_P = V_P$ (Planck volume). Also see [3] for a parallel derivation.

By substituting G with the corresponding Planck units, we get:

$$\boxed{G = \frac{1}{\delta_{ZP}\, t_P^2}} \quad , \qquad (2)$$

and substituting:

$$\boxed{G = \frac{1}{5.159x10^{96}\, kgm^{-3}\, (5.391x10^{-44}\, s)^2} = 6.670x10^{-11}\, m^3kg^{-1}s^{-2}} \quad , \qquad (3)$$

which is equal (to the rounded decimals) to the normal value of G ($6.673x10^{-11}$ $m^3kg^{-1}s^{-2}$) and demonstrates that G corresponds effectively to function (2), i.e., that **G is a quantum-function**.

In consequence, Newton's equation of gravitation adopts the following form by substituting G by eq. (2):

$$\boxed{F = \frac{1}{\delta_{ZP}\, t_P^2}\frac{m_1 m_2}{d^2}} \quad . \qquad (4)$$

In this equation, we call the above-mentioned independent right component "mass attraction" because it depends only on masses and their relative position. According to the "cause-effect principle", the left component (G) can be considered a "vacuum reaction" to gravitation, since matter is obviously the origin of any gravitation. (A similar vacuum reaction to accelerated matter is known as Davies-Unruh effect [9], [10] and demonstrates that the vacuum effectively reacts to the presence of matter. In addition, since in eq. (4), δ_{ZP} represents a vacuum mass-density equivalent, any reaction to gravitation represents a vacuum effect).

The first we observe in eq. (4) is that gravity is inversely proportional to vacuum energy (δ_{ZP}). In consequence, if we manipulated vacuum energy, we would in parallel be manipulating gravity. Further, if the QV did not exist, ZPR would be zero and according to (4), the gravitational force would be infinite while the opposite would occur if ZPR was infinite (gravity would be zero). In [11], we already mentioned a herewith-related case with regard to the extreme high temperature of the solar corona (up to $2x10^6$ °C) with respect to the photosphere or surface of the sun (only 5,500°C). Probably, the very dense solar photon stream produces "holes" in the fabric of spacetime, so that the ZPR emerges from QV and heats up the solar corona. In consequence, we predict that in the solar corona, gravity could be weaker than normal.

To understand the nature of the QV and why ZPR is able to reduce gravity, we make the following experiment of thought:

Imagine a flat universe (sheet) located in spacetime (our frame). Any spacetime radiation that crosses the sheet, will produce effects in the sheet, but will not remain there. Eventually the inhabitants of the sheet (flatlanders) will notice radiation effects, but will see no radiation. Analogous happens in our universe: vacuum radiation that crosses spacetime, produces Casimir-like effects, but cannot be seen nor detected because of its alien location.

Since the QV is a 6-D space [2] that surrounds 4-D spacetime completely due to its natural superior extension, spacetime matter is completely surrounded by ZPR. If an object is moving uniformly or at rest, ZPR will be the same on any surface. But as soon as the object accelerates, a Doppler-effect takes place (see also [3] for analogous explanation), so that ZPR becomes more dense in the direction of movement and less dense behind the object. This produces a higher ZPR pressure in the opposite direction of the movement, so that an effective "vacuum reaction" takes place, with the consequence that the initial acceleration is reduced. This effect is commonly known as '**inertia**' and per definition also somehow related to the much weaker Davies-Unruh effect. This means that the vacuum reacts to acceleration by opposing ZPR-borne inertia.

In the case of static bodies, ZPR produces a homogenous radiation pressure, so that no neat vacuum reaction nor inertia takes place. But if we managed to increase ZPR, according to (4), we would induce a vacuum reaction artificially and gravitation would therefore weaken in a parallel extent. In consequence, not only the acceleration is able to produce vacuum reaction (Davies-Unruh, inertia), but also any phenomenon that affects vacuum density.

This can be understood as a vacuum reaction opposite to the existing gravitational fields. Through Newton's equation of motion, $F=m\cdot a$, any body subjected to a field of force (in this case, gravitation), is also subjected to a potential acceleration toward attracting bodies. In consequence, the vacuum reaction will produce a reaction force via ZPR pressure that is opposed to the main direction of the corresponding gravitational fields, with the final result that, even bodies in a stiff gravitational system are subjected to a neat ZPR reaction force opposite to the direction of the field (although not to inertia, since they are not accelerated. In consequence, inertia and vacuum reaction to forces differ in stiff systems and are not exactly the same).

With the above model in mind, we interpret Podkletnov's famous "weak gravitation shielding experiment" as the obvious result of QV manipulation with the aid of electromagnetic (EM) fields, produced by a spinning superconductor. In effect, it seems that the fields induced by the superconductor and/or the coils used for levitation, did increase local ZPR, probably by uploading photons from EM fields to the QV. At that point, according to eq. (4), the higher ZPR should have reduced gravity, which was effectively observed by Podkletnov [4] and later also by Li [5] and several others.

This means that, it was not the superconductor itself that affected gravity, but the resulting EM fields. Probably, a flow of photons from spacetime to the QV takes place if strong EM fields interact mutually, e.g. by making one field rotate inside another as in [4]. The resulting 'friction' between both fields could provide the necessary energy to upload photons from spacetime fields to the QV, thus increasing local ZPR and reducing gravity according to eq. (4).

This principle suggests further that ZPR is not uniform in the universe, but that there could be many local phenomena that increase or decrease vacuum energy, making therefore respectively decrease or increase the local gravity. The universe can be in this sense understood at least as a giant gravity device via the QV. The expansion of the universe could be so, at least in part, interpreted as the result of ZPR repulsion on matter. Emerging ZPR in intergalactic space could explain intergalactic voids as well as the global dispersion of the universe.

To alter ZPR, we could use any kind of device or system, able to create a photon flow from a QV to spacetime and/or vice-versa. It seems at this stage easier to weaken gravity by mutually interfering, intense spinning EM fields, or by flooding tiny spaces with large amounts of photons [11], than to increase gravity by contrary means.

Anyhow, to increase gravity, it would be necessary to extract photons from QV. This may be the way that black holes, neutron stars and other dense objects do produce or absorb some radiation of this kind. By compressing supercold matter (Bose-Einstein condensate), instead of obtaining a fusion as in a conventional hydrogen bomb, as we have already suggested in [12], it should be able to create superdense matter. Arranging this matter in devices, it should be able to download photons from the QV in order to increase local gravity. A spacecraft equipped with a combination of several of the above-mentioned technologies, should be able to navigate without the need of any propulsion, at least in the proximities of celestial objects.

All the above-mentioned are methods dedicated to alter locally Gravitational Constant, 'G'. But the right component of eq. (4), which consists of "mass attraction", is the phenomenon that, in principle, induces a vacuum reaction (left component). This means that, apart from the vacuum reaction, masses attract mutually by some sort of a 'gravitational field' or 'spacetime geometry' as predicted by quantum field theories and general relativity, respectively. Therefore, another way to alter gravity would be by manipulating such fields or geometries - a theme that is out of the scope of this paper.

The novelty of eq. (4) is that it demonstrates that quantum gravity is not really subjected to any kind of constant, but that it is the result of the balance between gravitational attraction and vacuum repulsion on spacetime objects subjected to ZPR. Nature has probably provided this apparently complex system to allow a more stable universe that is not subject to unbiased and/or infinite field intensities and/or accelerations [1]. A universe without a QV component would probably be a very chaotic place if it had ever existed at all.

Following an analogous method as once Max Planck did when he derived Planck units from natural constants, and expressing Coulomb's Constant ($C = 8.988x10^9\ Nm^2/C^2$) in term of Planck units, in [2] we found the charge contained in a Planck volume (Planck charge):

$$q_P = \left(\frac{m_P\, c^2\, l_P}{C}\right)^{1/2} = \left(\frac{2.177x10^{-8}\, kg\left(2.998x10^{8}\, ms^{-1}\right)^2 1.616x10^{-35}\, m}{8.988x10^{9}\, Nm^2C^{-2}}\right)^{1/2} = 1.876x10^{-18}\, C \quad . (5)$$

This charge, divided by the charge of the electron ($1.6022x10^{-19}$ C), renders the amount of 11.71 leptons per Planck volume, corresponding to well-known virtual electron-positron pairs.

Since Coulomb's Constant is a function of vacuum permittivity (e_0) and/or permeability (μ_0), Planck charge can be interpreted as the charge of virtual pairs existing in vacuum. In consequence, any vacuum Planck volume contains 11.71 virtual leptons. Because 11.71 is a mean value, there can be 10, 11, 12, 13 etc. leptons at any time in a Planck volume, with 12 being the closest integer value.

These 12 leptons would correspond to 6 virtual pairs. It is known that any pair mutually annihilates and produces a gamma particle pair, which again mutually annihilate and renders an electron-positron pair in an almost immediate and endless sequence inside a confined space. At the very moment of their mutual interaction, 12 leptons/gamma particles fuse and represent the volume of 6 particles/strings. Since any string is considered to be one-dimensional and of the Planck length, 6 strings would fill up the space corresponding to a 6 dimensional Planck volume.

Since the above-mentioned 6 virtual pairs are contained in a Planck volume that according to Coulomb's Constant is a vacuum, it results that quantum vacuum is a **6-dimensional space**. (In fact, 6 one-dimensional strings concentrated in one spot have, per definition, the extension of a 6-dimensional Planck volume).

In [2], we also showed that, by comparing the fermion/photon ratio of spacetime and QV, it results that the QV contains 12 orders of magnitude more photons than spacetime. This enormous difference between these principally infinite macroscopic spaces, together with the above 6 dimensions that are found to exist in a QV, demonstrate that the QV and spacetime are different spaces, i.e. that they are not the same space.

Further, it is generally known from string theory that particles consist of 10-dimensional strings. These 10 dimensions correspond to 4 outside spacetime dimensions (the outer shell of the corresponding particle) and 6 curled dimensions inside string environment that are supposed to be a relic of the Big Bang, so that the original 10 dimensional particle that exploded at the beginning of the universe rendered a large amount of smaller particles that carry inside curled 6-dimensional (Kaluza-Klein) universes, each about the Planck size.

The 6 dimensions attributed to string environment coincide exactly with the number of 6 dimensions we found to exist in the QV. In consequence, the QV can be considered the medium in which strings are located, such that any Kaluza-Klein universe that surrounds a particular string would be connected to the QV through a path to the 'other side' that is not directly accessible through spacetime under normal conditions.

On the other hand, it is generally known that entangled particles do display a so-called 'non-locality'. That is the ability of particles like bosons to get synchronized at very large distances and is interpreted as the immediate interchange of information in a time that is zero for spacetime observers (see [13] for additional explanation).

With our model of strings that are surrounded by a 6-dimensional QV, particle non-locality can be easily understood as the ability of elementary particles to interact through the QV. As previously mentioned, each particle can be understood as consisting of an outer 4 dimensional shell (spacetime side) and an inner 6-dimensional string (QV side). While the outer shell would be responsible, e.g. for conventional photon interchange via spacetime, the inner shell would be responsible for mutual 'communication' that is known to exist in entangled particles. Since communication in 6-dimensional QV happens in a time frame beyond 4-dimensional spacetime, any QV interaction happens in a time that has no meaning for spacetime observers (i.e., the time needed by two entangled particles to interact results to be zero for us and for all of our spacetime devices).

As a result, our universe can be understood as consisting of two different spaces (spacetime and QV) linked together by elementary particles, such that the outer shell of any particle corresponds to the well known 4-dimensional particle, while the inner side is built up by a string surrounded by a local portion of the 6-dimensional QV. Generally speaking, any particle can be considered as being a small window to the QV.

Discussion

In the past, there has been no possibility to fuse these tiny windows in order to create a window that is large enough to be used, e.g. as a path to the 'other side', but the recent discovering of so-called "Bose-Einstein condensation" (BEC) makes it now possible. In fact, at a temperature of only one-billionth degree Kelvin, atoms turn into matter waves and acquire quantum properties. This means that BEC atoms behave like bosons and that they are able to overlap and to produce a single wave made out of all the small waves that constitute the corresponding atoms.

By fusing supercold atoms into a matter wave, what we would really be doing is eliminating the outer 4-dimensional shell of elementary particles, thus allowing the 6-dimensional QV to **emerge** and provide quantum properties to the wave, which are per definition usually confined inside the QV. By fusing a large amount of particles into one single wave, all Kaluza-Klein universes of the corresponding strings would also fuse, building a path to the QV, which could theoretically be enlarged with awesome potential. One possible use would be transportation through the great expanses of space.

A matter wave can be further understood as the "event horizon" of the border between spacetime and QV, resembling the event horizon of a black hole to some degree. In this sense, a matter wave (in opposition to a light wave) acts as a large window or path between spacetime and the QV. (Light can be understood as consisting of strings that are *not connected* to the QV and just simply "ride" upon spacetime fabrics. This 'simple' difference would mark the fundamental difference between matter and energy).

Theoretically, in order for a large BEC window to be used as a path for devices or probes to explore the other side, we ought to be able to create a BEC that is large enough for this purpose. Currently the BEC is very fragile and consists only of a diluted cloud of dispersed atoms. A technically useful BEC-window should consist of a large dense mass of BEC that is condensed solid matter, large enough, to provide sufficient space and dynamism for transportation through it. To achieve such a goal, we must improve the current techniques that allow creation of BEC (i.e., magnetic traps, laser light spotting, etc.) and use more powerful ones, such as matter blocks submerged in a circuit filled with a superfluid (e.g., He-4). Even if superfluids are not perfect BECs, a counter-flow of such fluids will produce a constant energy loss in matter blocks submerged in that flow, eventually until the BEC-temperature. In consequence, even with relatively simple technologies, it might be possible to produce large blocks of BEC, by just using the right technique.

Despite any difficulty that could exist in producing the above-mentioned technology, we are confident that they are in reach within the following 25 years or even earlier if these ideas are explored expeditiously.

Conclusions

Such ideas have been criticized about not using relativistic approaches to the corresponding equations presented. But this criticism is senseless because at the Planck level, there is **no relativity**. Max Planck derived his fundamental units directly from natural constants without using any relativistic approaches. Natural constants also provide information about the QV directly. In this sense, the speed of light is the exact value of Planck length divided by Planck time and vacuum energy corresponds exactly to Planck mass divided by Planck volume. Relativity probably takes place somewhere between the Planck world and normal level. But neither the Planck level nor the cosmological level seem to be subjected to any sort of relativity – they are fundamental and so are such concepts.

References

[1] Carlos Calvet, "Gravitation and inertia as a consequence of Quantum Vacuum Energy", Journal of Theoretics, Vol. 4, No. 2, April 2002 (preprint).

[2] Carlos Calvet, "About the quantum vacuum lepton/photon ratio", Journal of Theoretics, Vol. 4, No. 2, April 2002 (preprint).

[3] H.E. Puthoff, "The Energetic Vacuum: Implications For Energy Research, Speculations in Science and Technology", Vol. 13, No. 4, pp. 247-257 (1990).

[4] E.E. Podkletnov (Moscow Chem. Scientific. Ctr.), "Weak Gravitation Shielding Properties of Composite Bulk YBa2Cu3O7-x Superconductor Below 70°K under E.M. Field," Univ. Cincinnati Engineering, report # MSU-chem 95, abstract cond-mat/9701074, 19 pp. (1997).

[5] N. Li, D. Noever, T. Robertson, R. Koczor and W. Brantley, "Static Test for A Gravitational Force Coupled to Type II YBCO Superconductors", Physica C, Vol. 281, pp. 260-267 (1997) .

[6] S.K. Lamoreux, "Demonstration of the Casimir force in the 0.6 to 6 mm range", Phys. Rev. Lett., Vol. 78, No. 1, pp. 5-8 (1997).

[7] P.W. Milonni, R.J. Cook, & M.E. Goggin, "Radiation pressure from the vacuum: Physical interpretation of the Casimir force", Phys. Rev. A, Vol. 38, No. 3, pp. 1621-1623 (1988).

[8] H.G.B. Casimir, "On the attraction between two perfectly conducting plates", Proc. Kon. Ned. Akad. van Weten., Vol. 51, No. 7, pp. 793-796 (1948).

[9] P.C.W. Davies, "Scalar particle production in Schwarzschild and Rindler metrics", J. Phys. A, Vol. 8, p. 609 (1975).

[10] W.G. Unruh, "Notes on black-hole evaporation", Phys. Rev. D, Vol. 21, p. 2137 (1980).

[11] Carlos Calvet, “Effects and Evidence of the Background Field”, Journal of Theoretics, Vol.2, No.4, Aug 2000 (preprint).

[12] Carlos Calvet, “Evidence for the Existence of 5 Real Spatial Dimensions in Quantum Vacuum - Scale of Quantum Temperatures Below Zero Kelvin”, Journal of Theoretics, Vol.3, No.1, Feb. 2001 (preprint).

[13] Raymond Y. Chiao, Paul G. Kwiat, Aephraim M. Steinberg, "Quantum non-locality in Two-Photon Experiments at Berkeley" (International Workshop on Laser and Quantum Optics, Nathiagali, Pakistan, 9-14 July 1994) in Quantum and Semiclassical Optics 7, 259-78 (was preprint quant-ph/950101).